R. S. Williamson

On the Use of the Barometer on Surveys and

Reconnaissances

Being a Compendium, without Plates, of No. 15 of the Professional...

R. S. Williamson

On the Use of the Barometer on Surveys and Reconnaissances
Being a Compendium, without Plates, of No. 15 of the Professional...

ISBN/EAN: 9783744779289

Printed in Europe, USA, Canada, Australia, Japan

Cover: Foto ©berggeist007 / pixelio.de

More available books at **www.hansebooks.com**

ENGINEER DEPARTMENT, U. S. ARMY.

ON THE USE OF THE BAROMETER

ON

SURVEYS AND RECONNAISSANCES,

BEING

A COMPENDIUM, WITHOUT PLATES, OF No. 15

OF THE

PROFESSIONAL PAPERS OF THE CORPS OF ENGINEERS.

BY

LIEUT. COL. R. S. WILLIAMSON,

CORPS OF ENGINEERS, U. S. A.

WASHINGTON:
GOVERNMENT PRINTING OFFICE.
1878.

OFFICE OF THE CHIEF OF ENGINEERS,
    Washington, D. C., August 7, 1878.

SIR: Lieut. Col. R. S. Williamson, Corps of Engineers, has submitted to this office a compendium (without plates) of his paper "On the use of the Barometer," &c., Professional Papers, Corps of Engineers, No. 15.

This condensed work contains most of the information to be found in the larger one, with a few tables which are new, the result of further investigation of the subject.

As this compendium will be very useful and convenient to officers conducting barometric reconnaissances, I have the honor to recommend that it be printed at the Government Printing Office, and copies furnished upon the usual requisition.

    Very respectfully, your obedient servant,
                    H. G. WRIGHT,
                    Acting Chief of Engineers.

Hon. GEO. W. MCCRARY,
        Secretary of War.

Approved.
By order of the Secretary of War.
                    H. T. CROSBY,
                        Chief Clerk.

WAR DEPARTMENT,
    August 10, 1878.

SAN FRANCISCO, CAL.,
April 30th, 1878.

GENERAL : I have the honor to submit for your consideration a condensed copy of my work on meteorology and hypsometry, thinking that a small volume of this kind can easily be carried in the field and be usefully employed there, while the original work with its plates is not in a convenient form for that purpose. This little work contains most of the information to be found in the larger one, but I have added a few tables which are new, the result of further investigation of the subject.

In the concluding remarks I have compared the methods of treating meteorological observations with that of Prof. J. D. Whitney as described in his work entitled "Contributions to Barometric Hypsometry," and have shown conclusively that there are over forty per cent. more of maximum and mean errors by his method than by mine.

Very respectfully, your obedient servant,

R. S. WILLIAMSON,
Lieutenant Colonel of Engineers.

Brig. Gen. A. A. HUMPHREYS,
Chief of Engineers, U. S. A.

TABLE OF CONTENTS.

LIST OF TABLES.

INTRODUCTION.

To the large number of engineers, surveyors, and others, who are and will be engaged in developing the geography of this country, so large a portion of which is almost unknown or but partially explored, the best method of treating observations of the barometer and thermometer, so as to obtain the most reliable results in determining differences of altitude, is a matter of the first importance. It is well known that the mercurial cistern barometer is the best instrument for that purpose, for the reason that the spirit-level is out of the question, except within very limited areas, the length of time and amount of labor required for its proper use being far greater than can be devoted to the determination of the vertical element on an ordinary survey. The only instrument, that can be mentioned as at all to be compared with the cistern-barometer is the handy aneroid, the defects of which, however, as compared with the mercurial instrument, are so great as to preclude its being used as a substitute for the latter. Besides the fact that the aneroid is not susceptible of reading closer than one-hundredth of an inch, while the mercurial cistern can be read to one-thousandth, the great defect that it is liable at any time to change its zero, particularly in travelling, without there being any evidence to show that a change has occurred, makes the instrument entirely unreliable on a survey of any extent. The mercurial cistern barometer is, then, the only instrument that can be used with any satisfaction for hypsometrical purposes, and the following few pages will be devoted to show the best method of using it with its accompanying open air thermometer.

I may remark, in the first place, that whenever the readings of the barometer are referred to in the following pages, those of the barometer reduced to 32° Fahrenheit are meant. In the barometric formula of Laplace and others, a term has been introduced to take into account the effect of the expansion and contraction of the mercurial column by heat, in order to reduce the readings to what they would have been had the temperature of the instrument been always at the freezing point. But it is equally accurate and much more convenient to reduce each reading in the first place to the freezing point by the tables which have been prepared for the purpose. By adopting this course, the column so reduced, when plotted, shows the movements of a natural atmosphere, and their peculiarities can be studied with advantage; whereas the readings of the barometer not so reduced give so irregular a curve, the movements being masked by the ever-varying temperature of the instrument, that it is scarcely possible to discover any law guiding them, if such a law exists.

I also wish to point out that, unless special mention is made to the contrary, the formula used in the computations is the one found in Professional Papers of the Corps of Engineers, No. 15, only omitting the special correction for the moisture in the atmosphere. It is a translation of the formula of Plantamour. This formula differs from the one prepared by Guyot for the Smithsonian Institution, which is, in fact, the formula of Laplace, by a very small change in the barometric constant. Plantamour adopts the number 60,384.3, while Guyot gives 60,158.6. This slight change causes the difference of altitude to be greater by the former formula than by the latter by a little less than four feet for each thousand feet of difference of altitude.

I shall frequently have occasion to refer to the graphic representation of meteorological observations, which operation is called plotting. In order to represent the various movements of the atmosphere graphically, and in such a way that the value of the changes can be measured, it is necessary to attach scales to the drawings. In all cases to which I shall refer, the vertical scale is either a scale of inches of the barometric column, or of degrees of the thermometer. The horizontal scale is a scale of hours, or days, or month, as the case may be. But as I do not propose to illustrate this paper by such drawings, I shall endeavor to give my descriptions in such a way that my remarks will be easily understood without them.

OF THE HORARY AND ABNORMAL OSCILLATIONS OF THE BAROMETER.

A study of barometric observations, extended over a sufficient period of time, will reveal the existence of two distinct oscillations. These have been called respectively the horary and abnormal oscillations. The horary oscillation has a period of 24 hours. Within this period it presents, except during barometric storms, two distinct maxima and two minima, easily recognizable. The abnormal oscillation, on the other hand, is the result of a steady progressive movement of variable period, but usually it passes from one maximum to one minimum in from three to six days.

When a series of hourly observations of the barometer, taken during ten or more days, is plotted, there appears during each day a regular movement, more or less marked, and indicating two maxima and two minima in the twenty-four hours. If a table is made by taking separately the mean of the observations at the same hour of each day, thus obtaining twenty-four mean readings when the observations are taken hourly and uninterruptedly, and this mean table is plotted, the mean curve so developed shows this double oscillation very decidedly. If we were to make a grand mean by adding up the twenty-four mean hourly results, and dividing by twenty-four, and if we then subtract each mean result from the grand mean, we have a table in which some of the numbers would be greater and some less than the grand mean, and therefore some would be affected with a plus and some with a minus sign. This table can be used as a table of corrections, to be applied to the mean results at each hour separately, in order to reduce each reading to the mean value. This table would represent approximately a true table of horary corrections, but only approximately, unless the readings of the barometer at the beginning and end of the series happen to be the same, as will become apparent further on. Finally, if we apply these corrections to the original observations, we will have what has been called the "observations reduced." These, when plotted, show a wave-like movement in which no trace, or but a very slight trace, of the double horary oscillation appears. This curve represents very nearly the abnormal oscillation.

It is very apparent, from the study of such curves plotted from observations reduced to 32° Fahrenheit, that there are two separate forces in action, one producing an oscillation of regular period, and the other an irregular but slowly progressing movement of variable period. It is apparent, also, that during any twenty-four hours, the forces being in action together, the horary oscillation will appear more or less distorted or masked by the action of the abnormal movement. But inasmuch as the portion of the abnormal movement during one day often shows approximately a uniform rise or fall, the two coexisting movements can be easily separated. Let us suppose that the barometer during the day had been rising, and that it read two hundred and forty thousandths higher at the end of the day than at the beginning. If the abnormal oscillation during that day had been such that it could be represented by a right line, then that portion of the movement due to it would show a uniform rise for each hour. In the case I am supposing, the rise in one hour would be ten thousandths of an inch; in the first two hours twenty thousandths, etc. Now, if we were to apply a correction of ten thousandths to the reading of the barometer at one hour after the initial hour, one of twenty thousandths to that at two hours after the initial hour, etc., the table so resulting would be one representing the movement of the barometer freed (entirely in this case) from the effects of the abnormal movement.

But it is very rare that the abnormal movement during the twenty-four hours can be represented strictly by a right line. It is usually, when plotted, more or less curved, being a section of a sweeping curve which requires several days to pass from its maximum to its minimum. Moreover, if it so happens that the time when the abnormal wave reaches its maximum or its minimum in the middle of the day, a portion of that wave must be deeply concave and the remaining portion convex, and during that day the plotted observations, which represent the combination of the two movements, will show an irregular line different from the normal horary curve, though traces of that will probably be apparent. But it is exceedingly probable that during a series of ten days another day will be found in which a similar movement of the abnormal oscillation will occur, but of such a character that a portion of it will show a concave curve when the observations during a similar portion of the other day showed a convex one, and that the reverse will occur during the remaining portions of those two days. That is to say, the abnormal line will be a convex curve during one day and a concave one during the other. The combination of the observations during two such days would produce a curve approaching to a right line. It has been found from experience that observations taken for ten days, when treated in this way, generally produce a truthful and characteristic curve.

This method of treating observations, thereby eliminating the abnormal movements, has been called the "reduction to level." The difference between the reading of the barometer at the initial hour of two consecutive days is evidently the correction to level for twenty-four hours, which must be

2 U B

called minus when the barometer during the day has been rising, and plus in the reverse case, and the one twenty-fourth part of that is the correction to level for one hour. The correction to level for two hours is twice that for one hour, etc. When the correction to level for twenty-four hours is a multiple of twenty-four, the correction for each hour can be written down without difficulty, but when that is not the case it requires a little calculation to show at what hours the additions or subtractions shall be made. For example, if the correction to level for one hour were three-thousandths of an inch, the correction for the succeeding hours would be 6, 9, 12, etc., thousandths; but if it were a whole number of thousandths and a fraction, as, for example, three and fifteen twenty-fourths, as all the fractions less than half are to be thrown away, and all greater than half are to increase the whole number by one, we should have 4, 7, 11, 14, 18, 22, 25, etc., for the number of thousandths to be added or subtracted at 8, 9, 10, 11, 12, 1, 2, etc., hours, the barometric day beginning at 7 a. m. Table B of Professional Papers of the Corps of Engineers, No. 15, is intended to facilitate the calculations of the reduction to level by showing at what hours .001 is to be added on account of the fractional part of the correction for one hour. This table I have found convenient, but it is so simple of construction that any one can make it in a few minutes.

When the observations reduced to level are continued during several days, and are plotted, they show a series of curves occupying different parts of the paper, because the observations at the initial hour on different days will be different. When it is desirable, as is often the case in practice, to place them as nearly as possible in a horizontal row, it is best to subtract from each observation so reduced a certain number, which is the same for all the hours of one day, but differs in different days, so as to make the observations at 7 a. m. all alike. This second reduction, called "the reduction to second level," does not change in the slightest degree the character of the oscillation, but simply has the effect that, when plotted, the curves are found in a convenient part of the paper. It has been found best to adopt such a subtrahend for each day as will make the observations at 7 a. m. a little less than any one in the series. In California 29.500 is usually used near the sea-level, as the barometer seldom falls as low as that.

The following may then be given as a rule for obtaining a table of horary corrections of the barometer and one of the abnormal oscillations freed from the horary movement:

The observations (reduced to 32° F.) are to be copied in such a way that all at the same hour shall be placed in the same vertical column.

Each vertical column is to be added up and divided by the number of days in the series.

If hourly observations during the series are continuous, the grand mean is to be obtained by adding up the twenty-four mean results at each hour, and dividing the sum by twenty-four. If observations are not taken during the night hours, then an approximate grand mean is to be obtained by taking the mean of the observations at 7 a. m., 2 p. m., and 9 p. m.

Subtract the grand mean from the mean at each hour in succession, and we have a table of horary corrections, in which all those greater than the grand mean are to be called minus (−), and those less than the grand mean, plus (+).

The table of horary corrections being applied to the observations at 32° F., produces the table of observations reduced, which represents the abnormal oscillation.

It is of great importance, particularly when the series is not long, that the observations should be plotted, in order that any days' observations that are so erratic that they would evidently vitiate the result if they were combined with the others can be detected and rejected. The reason for this rejection is that we cannot expect the movements of the atmosphere during barometric storms to show its regular normal oscillation. Another advantage in plotting the observations is to afford the means of detecting errors in observations.

Any hour of the twenty-four may be taken as the initial hour of the barometric day, but it has been found expedient, for several reasons, to adopt 7 a. m. for that hour. Midnight is certainly an inconvenient hour, as observations are seldom taken at that time. When observations are taken but three times daily, the Smithsonian hours of 7 a. m., 2 p. m., and 9 p. m. are almost invariably adopted in this country. Full observations taken during all of the twenty-four hours are of great value for certain scientific purposes, particularly in deducing the laws which cause the horary oscillation to be different in different latitudes, altitudes, and climates. But as barometric observations for hypsometrical purposes are seldom taken during the night, the practical engineer or surveyor seldom cares for them later than 9 p. m., even though he does not obtain the night maximum and morning minimum, the first of which occurs not far from midnight, and the latter or second between four and five in the morning.

When the series is a short one and the horary table is made with the hope of obtaining a type curve, or when it is suspected that the observations, as usually treated, will not give a result fully freed from the abnormal movement, another horary table may be made by treating the observations in the same way, but adopting 7 p. m. as the initial hour instead of 7 a. m. If the series be of ten days' duration, the new table with the new initial hour will produce a nine days' series. A combination of the two mean tables is more likely to produce a good horary curve than either one separately. This would be naturally the case, for, from the very principle of the reduction to level, it is assumed that

the portion of the abnormal wave during each day is a right
line, whereas it is only approximately so, and is in reality a
curve more or less convex or concave. Now, by adopting
another initial hour an additional horary table can be made,
and it is more than likely that by the combination of the
two the abnormal movement is more thoroughly eliminated

When the series of hourly observations is not continuous
as, for instance, when the night observations are not taken,
we cannot use the method just given, but we must adopt
some method of obtaining approximately a grand mean, and
it has been found that the mean of 7 a. m., 2 p. m., and 9
p. m., gives a close approximation to the mean of the
twenty-four hourly observations. It is important to know
how close the agreement is. Although the number of stations where I have been able to collect observations is quite
small, still they have been sufficient to prove, not only that
the difference in obtaining the daily mean pressure by the
two methods is not great, but that it varies by a regular
law during the months of the year. This is shown by the
following table :

TABLE I.—*Showing the difference between the mean barometric pressure in the different months, as obtained from the mean of twenty-four hourly observations and observations at 7 a. m., 2 p. m., and 9 p. m., the former being assumed as the standard.*

Stations.	January.	February.	March.	April.	May.	June.	July.	August.	September.	October.	November.	December.	Yearly mean.
Astoria	+.003	−.003	+.003	−.005	−.006	−.004	−.003	−.003	−.001	.000	+.001	+.004	−.001
San Francisco	+.006	+.006	+.002	−.002	−.004	−.004	−.004	−.003	−.002	+.002	−.002	+.010	−.001
Sacramento	+.002	−.002	−.002	−.006	−.006	−.003	−.001	−.001	+.002	−.001	+.003	+.002	.000
Placerville	+.003						−.006	−.001					
Strawberry Valley	+.004					−.004	−.004	−.003					
Hope Valley	+.003					−.002	−.002	−.003	−.002				
Carson City						−.003	−.003						
Toronto	+.005	+.001	−.001	−.007	−.004	−.005	−.004	−.004	−.003	+.001	+.003	+.004	−.001
Girard College	+.005	+.003	+.002	−.002	−.003	−.004	−.004	−.001	−.002	+.001	+.004	−.006	.000
Geneva	+.001	.000	−.002	−.002	−.001	−.002	−.001	−.003	−.012	−.001	+.001	+.001	−.001
Grand Saint Bernard	+.001	.000	−.001	−.002	−.002	−.001	−.001	−.002	−.001	−.001	−.001	+.001	−.001
Greenwich	+.003	+.001	−.003	−.003	+.001	−.001	−.001	−.002	−.001	−.001	+.002	.000	.000
Sun	+.036	+.006	+.002	.029	.026	.021	.037	.026	.012	.000	+.011	+.016	.005
Mean	+.003	+.001	.000	−.003	−.003	−.003	−.003	−.003	−.001	.000	+.001	+.002	−.001

By examining the table, it will be seen that out of the 107 differences in the monthly results by the two methods there is one amounting to ten thousandths of an inch of the barometric column in the stormy month of December, one of seven thousandths, six of six thousandths, four of five thousandths, sixteen of four thousandths, and all the rest are less than that amount. In fact, out of the 107 results there are only twelve in which the differences are so great as five thousandths. The mean results show that the mean of observations at 7 a. m., 2 p. m., and 9 p. m. is *less* than the mean of twenty-four hourly observations in the months of November, December, January, and February, the January results giving a difference of three thousandths of an inch, and that in the midsummer months it is *greater* by the same amount. In March and October there is no difference. The yearly mean difference by the two methods is less than one-thousandth of an inch. The above table can be used as a table of corrections to be applied to observations taken in any month in order to reduce them to the yearly mean.

This table has been deduced from monthly means, and it is not to be supposed that observations taken in a single day, or even a series of a few days' duration, will afford so close an accord by the two methods. There is, however, a method, when the observations are taken at the Smithsonian hours, which affords a very good value for the daily mean, as compared with the mean of twenty-four observations in a day beginning at 7 a. m. It is to take the mean of 7 a. m., 2 p. m., 9 p. m., and 7 a. m. of the succeeding day and apply to it a correction, so as to reduce it to what it would have been had the observations been taken strictly at eight hours apart. From 7 a. m. to 2 p. m. is an interval of but seven hours, and from 7 a. m. to 9 p. m. is 14 hours. Hence the sum of these four observations is too small by the rise, or too great by the fall, between 2 p. m. and 3 p. m., together with the rise or fall between 9 p. m. and 11 p. m.; that is to say, during three hours. But 3 hours is one-eighth of a day, in which day the barometer had an average rise or fall measured by the amount which has been called "the correction to level for twenty-four hours," and which is the difference in the readings of the barometer at 7 a. m. on one day and 7 a. m. on the next. Therefore one-eighth of that amount should be applied as a correction to the sum of the four observations in question, so as to increase that sum when the barometer for the day had been rising and diminish it when falling. It must be borne in mind, however, that this correction is to be applied to the "observations reduced," or, in other words, to the observations after the horary correction had been applied.

We may therefore have the following rule for obtaining the mean daily pressure from four such "observations reduced," the barometric day commencing at 7 a. m. Take the sum of the observations so reduced at 7 a. m., 2 p. m., 9 p. m., and 7 a. m. the next morning and apply to it one-eighth of the difference between the observations at the beginning of the two consecutive days, calling the difference *plus* (+) when the barometer for the day had been rising and *minus* (−) in the reverse case, then divide the result by four, and we have the required daily mean.

As it is desirable to have a definite idea of the difference between the daily barometric means as computed from twenty-four hourly observations and that given by the last described method, and also between the former and that from observations taken at 7 a. m., 2 p. m., and 9 p. m., I present a table from ten days' observations at Sacramento, Placerville, Strawberry Valley, and Hope Valley, taken during August, 1860. The upper line of each group gives the difference between the daily mean as calculated from twenty-four hourly observations and the mean of 7 a. m., 2 p. m., 9 p. m., and 7 a. m. the next morning, corrected as above described. The lower line gives the difference between the first and the mean of 7 a. m., 2 p. m., and 9 p. m. It will be seen that the amount of variation from the mean of twenty-four hourly observations is nearly three times greater by the last mentioned method than by the first.

TABLE II.—*Showing the difference between the daily mean barometric pressure, as obtained from the mean of twenty-four hourly observations and from corrected observations at 7 a. m., 2 p. m., 9 p. m., and 7 a. m. the next morning, and also between the first and those obtained from observations at 7 a. m., 2 p. m., and 9 p. m., the first being assumed as the standard.*

August, 1860. STATIONS.	11th.	12th.	13th.	14th.	15th.	16th.	17th.	18th.	19th.	20th.	Sum.
Sacramento	.007	.002	.001	.003	.003	.003	.001	.001	.005	.002	.027
	.010	.008	.023	.001	.004	.006	.001	.020	.012	.000	.191
Placerville	.014	.001	.001	.001	.005	.005	.000	.001	.004	.004	.039
	.018	.014	.016	.002	.012	.002	.002	.001	.017	.004	.096
Strawberry Valley	.008	.001	.004	.000	.006	.000	.010	.011	.005	.005	.050
	.022	.008	.009	.006	.014	.005	.012	.010	.012	.006	.110
Hope Valley	.008	.004	.001	.003	.002	.004	.003	.006	.003	.004	.038
	.030	.009	.007	.012	.010	.005	.007	.005	.008	.005	.104

In almost every instance it is shown that the second method gives results much nearer the mean of twenty-four hourly observations than the third, which is a simple mean of the observations at 7 a. m., 2 p. m., and 9 p. m. Taking the mean of twenty-four hourly observations as the standard, and taking the difference between this standard and the results given by the other two methods, we find that the sum of these differences in a ten days' series by the method of four observations so reduced is only 38 per cent. of the corresponding amount obtained from a simple mean of the observations taken at 7 a. m., 2 p. m., and 9 p. m.; and the maximum errors are in proportion, they being at Sacramento .007 in. and .032 in., and at Hope Valley .008 in. and .030 in. As this method of obtaining the barometric mean involves very little additional trouble after the observations have been actually taken, it appears to me worthy of being adopted. For, in the field, we cannot tell when the atmospheric conditions which cause the difference between the two methods are in operation, and when the maximum difference will occur: hence the results are apt to be considered untrustworthy up to the limit of the maximum error. On the other hand, if that maximum error can be reduced two-thirds or one-half, the results can surely be relied upon within the smaller limit.

Although observations at the Smithsonian hours have been shown to give a close approximation to the mean of twenty-four hourly observations, still other hours have been adopted that give, also, very good results. The Coast Survey adopted long ago the mean of 6 a. m., noon, and 6 p. m. as a good barometric mean, though latterly they have adopted the hours of 7 a. m., 2 p. m., and 9 p. m. I have obtained from Louis Wilson, tidal observer at Astoria, Oregon, the following table, which explains itself:

TABLE III.—*Showing the difference between the monthly mean barometric pressure, as computed from observations at 7 a. m., 2 p. m., and 9 p. m., and 6 a. m., noon, and 6 p. m., the former being used as a standard.*

	January.	February.	March.	April.	May.	June.	July.	August.	September.	October.	November.	December.	Mean.
1869	+ .002	− .006	.000	+ .004	+ .001	+ .006	+ .005	+ .001	+ .005	− .001	− .002	.000	+ .002
1870	− .001	+ .001	+ .001	+ .005	− .004	+ .006	+ .006	+ .005	+ .002	.000	− .003	− .001	+ .001
1871	+ .001	− .003	− .004	.000	+ .008	+ .006	+ .003	+ .005	+ .001	− .001	+ .002	− .008	+ .001
Mean	+ .001	− .004	− .001	+ .003	+ .003	+ .006	+ .004	+ .005	+ .003	− .001	− .001	− .003	+ .001

This table shows nearly as good results as from observations at the Smithsonian hours, but as the latter hours have been so universally adopted, and it is very desirable to have uniformity in the hours of observation, so that comparisons can be easily made. I would recommend that the Smithsonian hours be adhered to in future observations where they are taken but three times a day.

Having explained how the horary correction of the barometer can be obtained from a short series of observation, I now wish to point out certain facts concerning this oscillation. It has been found that when the stations are near the sea level, the curves for each month at different localities are of the same character, the critical hours occurring at the same times, but varying in range or amplitude, the warmest localities giving the largest curves. Hence, as a general rule, the curves are smaller as the latitude increases. But in the same latitude and climate the curves for the different months are different. They vary in the hours of maxima and minima, and also in the amplitude of the oscillation. While the hour of the morning maximum does not materially vary during the different months, that of the afternoon varies with the seasons, being usually between 2 and 3 p. m. in midwinter and between 5 and 6 p. m. in midsummer. The consequence of this is, that if the tables representing the hourly observations taken in January and July are subtracted the one from the other and this difference plotted, it gives a curve nearly as great as is produced from either set of observations. But as soon as the element of altitude enters into consideration, the curve changes materially, and according to a law which has not yet been discovered. As a general rule, the curves for high altitudes are quite small. At the Grand Saint Bernard, that portion of the midsummer curve for the hours when the sun is above the horizon is exceedingly minute, while the night portion of the curve presents an oscillation of about 0.010 inch. Near the summit of the Sierra Nevada in July and August, the morning maximum is at 7 a. m., while in the valley below it is at 11 a. m. There is no similarity between the Grand Saint Bernard curves and those of the Sierra Nevada, though the altitudes of the two stations do not differ materially. If we had a series of stations one thousand feet apart, vertically, from the sea level to the summit of the mountain, we would find that the curves at all the stations would be different.

The amplitude of this oscillation in the temperate zone usually varies from 0.040 in. to 0.080 in. Near the equator the oscillation is greater, amounting to nearly 0.120 in., and the abnormal oscillation being there very small, the horary oscillation is so regular, that the hour of the day can be ascertained, at least approximately, from the reading of the barometer. But the abnormal oscillation seems to increase with the latitude, while the horary movement becomes less, and in high latitudes the latter is so masked by the former, that a long series of observations is required to obtain a reliable horary curve.

From the above facts it becomes apparent that the effects of this horary oscillation ought to be neutralized in some way. The computed differences of altitude from observations taken at different hours are different on account of the oscillations at the lower and upper stations being so entirely different. Now, as a change of 0.001 in. in the barometer at one station will affect the result about a foot, unless a corresponding change occurs at the other station, it is apparent that we should correct the observations *before* they are used in the determination of altitudes, so as to eliminate the effect of the horary movement. The following general conclusions are given in Professional Papers of the Corps of Engineers, No. 15, together with a large number of horary tables and curves:

1st. As the value of the principal term of the barometric

formula depends upon the *difference* between the readings of the barometers at an upper and lower station, and as the horary oscillation of the barometer is quite different at the two stations when the difference of altitude is at all considerable, and as its amount is often sufficient to cause considerable error in hypsometrical calculations if neglected, even when the observations at the two stations are simultaneous, it is important to eliminate it as far as practicable.

2d. As the horary curves and tables for any two days, even in a short series, are not identical, the best way to eliminate the effect of this oscillation is to use the mean of observations taken at short intervals, as, for instance, hourly, for one *day*, or for a number of *whole days*, the day commencing at any convenient hour.

3d. When this is impracticable, and when the horary tables for the station and month are previously known, and the observations are for a portion of a day only, or for portions of several days, the horary correction should be applied to them before they are used in estimating differences of altitudes.

4th. When the horary tables for one or both stations are unknown, and hourly observations cannot be taken, the aim should be to obtain the nearest approximation to a daily mean. For this purpose, the mean of observations taken at 7 a. m., 2 p. m., and 9 p. m., or of 6 a. m., 2 p. m., and 10 p. m., or 6 a. m., noon, and 6 p. m. have been found to afford quite good results.

ON THE VARIATIONS IN TEMPERATURE.

While the horary barometric oscillation, when freed from the abnormal movement, does not vary much from day to day at the same station during a short series, it is very different with the corresponding thermometric oscillation. In the one case it is small as compared with the abnormal one, and so nearly uniform in character that a mean of a few days' observations, properly treated, will give a characteristic horary table and curve for that station and month; and, by elimination, the abnormal wave can be represented. In the case of the temperature, the horary movement is very large as compared with the other, and varies so much from day to day that no characteristic horary table can be used in eliminating this movement, and obtaining an abnormal thermometric wave; though the curve is a simple one, having but one maximum and one minimum in 24 hours, still the range, or vertical amplitude, may be several times as great in one day as in another during a series of ten days. The consequence is that the method of separating the two movements, which we have found practicable with the barometric observations, is not applicable to those with the thermometer.

While the barometer gives us a measure of the weight of the whole column of air over the place of observation, the thermometer is local in its character and affected by every puff of wind that blows over it. It is true that there is one paramount influence which produces a horary thermometric oscillation with one decided maximum and one minimum, the former usually occurring between 2 and 4 p. m. and the latter about one hour before sunrise; but the amount of variation during the day is greatly modified by many accidental causes, such as the clearness or cloudiness of the atmosphere, the direction and force of the wind, the rapidity or slowness of the evaporation or condensation of aqueous vapor, and many other local meteorological phenomena. For these reasons the amount of this oscillation must vary greatly from day to day, and this experience shows us to be the case.

If, in a series of ten days' observations, the horary thermometric oscillations are plotted, it will almost always be found, in temperate latitudes, that the vertical range in the curve for some one day will be twice as great as for another in that short series, and it is not unusual to find that the difference is three and even four times as great. It is this great difference in range from day to day which prevents us from using advantageously a mean horary thermometric table for hypsometrical purposes. The same reason which makes it necessary to eliminate the effects of the horary oscillation of the barometer applies with still greater force to the varying temperature. It is principally by means of the pressure at the two stations, in connection with the corresponding temperature, that we are to obtain the difference of altitude between them. If the change in temperature from hour to hour caused a proper corresponding change in the height of the barometer, we could disregard the effects of the horary oscillations altogether and use the observed pressure and temperature at the two stations. But this is not so. When we take twenty-four hourly observations of the barometer and thermometer at stations of considerable difference of altitude, and estimate the vertical distance between these stations by the formula, using successively each pair of corresponding observations, we have a series of twenty-four numbers far from being alike. Again, when we do the same with the next two sets of twenty-four observations taken during the next day, we have another series differing from the first, and it would be very materially different if the horary thermometric oscillations for the two days are quite different, as they are apt to be. For these reasons it is evident that the horary oscillations of both the barometer and thermometer must be eliminated *before* the observations can be properly used in estimating differences of altitude. Yet I believe it is a common practice with computers to use the observed air temperature.

When hourly observations of the thermometer are taken, and the monthly mean for the different months are obtained, it has been found that the range of the horary thermometric oscillation varies from month to month, the same being greatest in the hottest months and least in the coldest.

The difference in these ranges seems to be greatest when the difference between the mean temperatures of the hottest and coldest month is greatest. The range is usually greater in arid districts than in the more humid ones near the sea. It has been found from observations at stations varying in altitude from the sea level to the summit of the Sierra Nevada, that in August the range at Sacramento, near sea level, was 17 degrees; at Placerville, about 2,000 feet high, it was 31 degrees; at Strawberry Valley, about 5,700 feet high, it was 33 degrees; and at Hope Valley, 7,000 feet high, it was 17 degrees. In January, at the same stations, the range was 11 degrees at Sacramento, 17½ degrees at Placerville, 16½ degrees at Strawberry Valley, and 17 degrees at Hope Valley. It therefore by no means follows that the range of this oscillation diminishes with the altitude, though it is doubtless small at exceedingly high places.

It is evident from the preceding remarks that the question of how best to obtain the daily mean temperature, in order to secure good hypsometrical results, is of great importance. Fortunately, this is not difficult. With the barometer the daily mean can only be ascertained from observations taken during that day at the precise locality, or approximately so, by applying a horary correction But with the thermometer the case is different. As the temperature during a day is nearly the same over a large area in a level country, observations can usually be taken in the field at 7 a. m., 2 p. m., and 9 p. m., and a good mean value to the temperature of the day thus obtained, although the party has been in motion. If the party has been in an uneven or mountainous region, then the only way is to assume that the mean daily temperature varies three degrees with each thousand feet of difference of altitude. I am aware

3 U B

that this rule gives but a very rude approximation to the truth, but, except when the change in altitude from camp to camp is very great, the error from adopting it will be small.

The following table gives a comparison between monthly mean temperatures obtained by twenty-four hourly observations and the mean of those taken at 7 a. m., 2 p. m., and 9 p. m. :

TABLE IV.—*Showing the difference between the mean temperature in the different months, as obtained from the mean of twenty-four hourly observations and those taken at 7 a. m., 2 p. m., and 9 p. m., the former being assumed as the standard.*

Stations.	January.	February.	March.	April.	May.	June.	July.	August.	September.	October.	November.	December.	Mean.
Girard College	0.26	0.17	0.15	0.53	0.57	0.77	0.61	0.41	0.35	0.33	0.28	0.29	0.39
Frankfort Arsenal	0.39	0.34	0.36	0.29	0.29	1.00	1.02	0.79	0.65	0.74	0.34	0.50	0.61
Toronto (Lefroy)	0.40	0.01	0.09	0.34	0.40	0.73	0.96	0.55	0.16	0.28	0.29	0.19	0.43
Toronto (Dove)	0.38	1.11	0.32	0.45	0.56	0.52	0.54	0.16	0.70	0.77	0.32	0.05	0.41
Rio Janeiro	0.33	0.50	0.43	0.27	0.41	0.27	0.47	0.32	0.32	0.32	0.34	0.36	0.36
Trevandrum	0.36	0.39	0.19	0.36	0.34	0.57	0.23	0.27	0.29	0.32	0.27	0.32	0.32
Madras	0.26	—	0.49	0.53	0.45	0.59	0.58	0.44	0.41	0.42	0.33	0.14	0.43
Bombay	0.14	0.02	0.09	0.05	0.20	0.16	0.30	0.13	0.07	0.14	0.18	0.09	0.11
Barnaul	0.86	0.43	0.56	0.74	1.31	1.49	1.54	1.13	0.63	0.47	0.59	0.50	0.54
Plymouth	0.14	0.27	0.27	0.45	0.08	0.59	0.70	0.41	0.36	0.29	0.29	0.11	0.36
Greenwich	0.20	0.30	0.26	0.60	0.90	1.20	0.70	0.70	0.40	0.20	0.30	0.10	0.30
Leith (Scotland)	0.14	0.14	0.18	0.29	0.27	0.36	0.43	0.27	0.32	0.16	0.41	0.18	0.36
Rome (Italy)	0.34	0.13	0.02	0.07	0.05	0.11	0.29	0.07	0.13	0.34	0.27	0.27	0.11
Padua (Italy)	0.16	—	+0.02	0.16	0.88	0.84	1.24	0.67	0.63	0.29	0.36	0.43	0.47
Kremsmünster	0.27	0.49	0.29	0.57	0.92	1.17	0.70	0.72	0.52	0.63	0.38	0.36	0.56
Prag (Bohemia)	0.29	0.34	0.36	0.34	0.36	0.22	0.99	0.67	0.70	0.59	0.45	0.34	0.47
Brussels (Belgium)	0.27	0.31	0.38	0.45	0.65	0.90	0.63	0.61	0.45	0.27	0.11	0.09	0.43
Mühlhausen	0.29	0.45	0.34	0.52	0.85	1.35	1.06	0.77	1.15	0.59	0.13	0.13	0.65
Halle	0.35	—	0.41	0.77	1.60	1.57	1.46	1.10	0.79	0.65	0.45	0.22	0.81
Göttingen	0.34	0.20	0.20	0.41	0.99	1.10	1.19	1.06	0.45	0.52	0.27	0.95	0.63
Berlin	0.20	0.34	0.41	0.43	1.26	1.53	0.81	0.34	0.58	0.88	0.38	0.20	0.56
Salzuflen	0.22	0.40	0.38	0.51	0.84	0.94	0.67	0.61	0.56	0.49	0.54	0.11	0.54
Stettin	0.29	0.29	0.29	0.43	0.85	0.81	0.90	0.56	0.31	0.45	0.18	0.31	0.47
Apenrade	0.18	0.47	0.38	0.36	0.93	1.06	0.74	0.70	0.52	0.20	0.25	0.18	0.59
Katharineuburg	0.36	0.31	0.40	0.47	0.99	1.33	0.97	0.47	0.40	0.55	0.20	0.40	0.45
Petersburg	0.22	0.11	0.29	0.34	0.72	0.65	0.49	0.49	0.38	0.20	0.04	0.14	0.36
Helsingfors	0.34	0.54	0.58	0.90	1.15	1.69	1.42	1.12	0.70	0.65	0.47	0.38	0.83
Christiania	0.27	0.16	0.11	0.52	1.12	1.15	0.88	0.72	0.52	0.13	0.31	0.16	0.54
Drontheim	0.13	0.16	0.27	0.70	0.64	1.15	0.92	0.94	0.58	0.34	0.27	0.16	0.54
Straits of Karn	0.07	0.04	0.56	0.90	0.65	0.49	0.38	0.11	0.13	0.04	0.16	0.09	0.30
Matoshkin Shar	+0.20	0.04	0.29	0.28	0.81	0.58	0.47	0.29	0.31	+0.34	+0.13	0.04	0.20
Sum	6.03	8.77	9.02	14.03	23.64	26.43	23.43	17.46	14.77	11.64	9.87	7.91	14.46
Mean	0.26	0.28	0.31	0.45	0.76	0.85	0.76	0.56	0.48	0.38	0.59	0.23	0.47

It will be seen that the mean temperature of 7 a. m., 2 p. m.. and 9 p. m., at almost every station, gives a result too great, in every month, as compared with the mean of twenty-four hourly observations: but the difference is not great, seldom exceeding in any one month one and one-half degrees. The mean results show that the mean temperature thus ob tained by the two methods most nearly agrees in December, where the difference is less than one-quarter of a degree, an d that the difference is greatest in June. From December to Jun e the difference increases with much uniformity, and from June to December it decreases in the same manner; so that if the table were plotted it would show a smooth curve. This table can be used as a table of corrections, to be applied to observations taken at 7 a. m., 2 p. m., and 9 p. m., in order to reduce them to the mean of twenty-four hourly observations.

I next present a table of comparison between the means of thermometric observations taken at 7 a. m., 2 p. m., and 9 p. m. and that of those taken at 6 a. m., noon, and 6 p. m. They were furnished me by Louis Wilson. tidal observer at Astoria, Oreg., and are for three years.

TABLE V.—*Showing the difference between the monthly mean temperature, as computed from observations at 7 a. m., 2 p. m., and 9 p. m., and 6 a. m., noon, and 6 p. m., the former being used as a standard.*

	January.	February.	March.	April.	May.	June.	July.	August.	September.	October.	November.	December.	Mean.
1869	— 0.0	+ 0.1	— 0.5	— 0.4	— 0.5	— 0.8	— 0.8	— 0.6	— 0.4	— 0.2	— 0.2	— 0.1	— 0.4
1870	— 0.1	— 0.1	— 0.1	— 0.5	— 0.1	– 0.5	— 0.7	— 0.2	— 0.2	— 0.1	— 0.2	— 0.1	— 0.2
1871	+ 0.2	— 0.1	— 0.2	— 0.4	— 0.2	— 0.7	— 1.0	— 0.9	— 0.0	— 0.1	— 0.0	— 0.0	— 0.3
Mean	— 0.0	— 0.0	— 0.3	— 0.4	— 0.3	— 0.7	— 0.8	— 0.6	— 0.2	— 0.1	— 0.1	— 0.1	— 0.3

It appears from this table that while the winter months give results which agree almost exactly by the two methods. the results in the summer months differ by about three-quarters of a degree, the mean of 6 a. m., noon, and 6 p. m. being the greater. As in those months the mean of 7 a. m.. 2 p. m., and 9 p. m. gives results too great by about that amount as compared with the mean of twenty-four hourly observations, it follows that the other method must be considered decidedly inferior.

From an examination of the monthly mean temperatures from year to year at Geneva and the Grand St. Bernard it has been found that when at one station the monthly mean temperature departs considerably from that determined by the mean of a long series, it is not local, for the same departure is found at the other station.

OF HYPSOMETRICAL RESULTS FROM DAILY MEANS.

It is presumed that the reader understands the barometric formula, and therefore but few remarks concerning it are necessary here. The most important parts of it are the pressure and temperature terms. The first consists of a constant, multiplied by the *difference* between the logarithms of two numbers, which are the readings of the barometer at a lower and an upper station. The second term is the product of the former divided by 900 and multiplied by the *sum* of two numbers, which sum, when the Fahrenheit scale is used, is the sum of the readings of the open-air thermometer at the two stations, diminished by 64, which is twice the temperature at the freezing-point.

The other terms of the formula are usually comparatively small, though by no means to be disregarded in the computations. When the formula is applied to observations taken for some time at the same two stations (in which case the true difference of altitude between them will, of course, be constant), the values of these terms should be constant.

With such a series, the mean readings of the barometers and thermometers can be used to compute the mean difference of altitude between them. When this is done, and the value of those small terms is once determined, if separate computations are made to obtain the difference of altitude from each day's observations, the same value for the sum of these small terms can be used, thus avoiding much useless labor in computing. When the pairs of stations are different, separate calculations must be made to obtain the values of these terms, which can be conveniently done by the aid of the tables in the appendix to Professional Papers of the Corps of Engineers, No. 15, before mentioned.

If computations are made from the daily means of observations during every day in a month at the same two stations, each result will vary more or less from the monthly mean. If a table of wanderings from the mean is made, some of the numbers will, of course, be greater and some less than the mean, and should be written, some with a plus (+) and some with a minus (−) sign. If all these errors or wanderings are added together, without regard to sign, and divided by the number of days in the month, the resulting number will give the mean error; and it may be confidently assumed that if similar observations at those two stations are taken during the same month of another year, very similar results will be obtained. But it can be shown from a large number of computed differences of altitude from daily means that the amounts of the maximum and mean errors in different months at the same two stations vary considerably, being least in mid-summer and greatest in mid-winter.

As a general rule, when the difference of altitude between the two stations is at all considerable, the amounts of the maximum and mean errors vary with the difference of altitude. This is quite natural, for, in the first place, when the difference of altitude is great the horizontal distance between them must also be comparatively great, and the errors are apt to be greater than when the stations are close together; but, in the second place, the temperature term of the formula is the product of the approximate difference of altitude given by the pressure term multiplied by a variable depending upon the temperature. Now, the value of this variable being different from day to day, the wanderings from the mean value of the temperature term are inevitable, and must be proportional to the value of this variable, as well as to the difference of altitude. This last cause of error, however, does not materially affect the result when the difference of altitude is quite small, in which case the errors are apt to vary with the horizontal distance between the stations.

The following tables of maximum and mean errors in computing differences of altitude from daily means are now presented. They are the result of great labor, as every two corresponding numbers are the result of as many different computations as there were days in the month: but they give a clear idea of the probable amount of error to which such results are liable.

TABLE VI.—*Consolidated table of maximum errors in computing differences of altitude from daily barometric and thermometric means.*

	Difference of altitude.	Distance.	Maximum errors.											
			January.	February.	March.	April.	May.	June.	July.	August.	September.	October.	November.	December.
	Feet.	Miles.	Feet.	Feet.	Feet.	Feet.	Feet.	Feet.	Feet.	Feet.	Feet.	Feet.	Feet.	Feet.
Geneva and Grand Saint Bernard, 1862..	6,792	60	136	120	192	79	75	91	128	55	106	114	154	223
Geneva and Grand Saint Bernard, 1863..	6,792	60	216	85	152	91	112	108	62	163	145	194	116	116
San Francisco and Sacramento	81	74						39	39	34	178			
San Francisco and Hope Valley	7,072	148												
Sacramento and Fort Churchill	4,238	124	203	267	241	182	161	129	92			230	262	275
Sacramento and Hope Valley	6,362	45	230						109	109	171			
Sacramento and Hope, with intermediate stations	6,991	57	181						52	84				
Placerville and Strawberry Valley	3,742	35	71						69	46				
Strawberry Valley and Hope Valley	1,365	12	93						43	41				
Sacramento and Placerville	1,844	40	34						41	30				
Sacramento and Fort Crook	3,292	179	290	215	192	205	88	177		129	91		193	217
Fort Crook and Hope Valley	3,670	195								86				
Hope Valley and Aurora	403	64	118											
Carson City and Hope Valley	2,335	30							32					
Sacramento and Aurora	7,365	142	244	154						149		183	400	182

	Difference of altitude.	Distance.	January.	February.	March.	April.	May.	June.	July.	August.	September.	October.	November.	December.
	Feet.	Miles.	Feet.	Feet.	Feet.	Feet.	Feet.	Feet.	Feet.	Feet.	Feet.	Feet.	Feet.	Feet.
Geneva and Grand Saint Bernard, 1862	6,792	60	41	49	51	26	34	32	25	18	36	37	55	59
Geneva and Grand Saint Bernard, 1863	6,792	60	59	32	44	26	31	33	15	35	32	51	41	41
San Francisco and Sacramento	81	74	...	...	...	...	...	...	13	11	x	91	76	143
San Francisco and Hope Valley	7,072	147	...	...	...	70	58	45	39	...	72	...	...	...
Sacramento and Fort Churchill	4,258	198	72	110	101	...	...	...	39	37	...	...	...	...
Sacramento and Hope Valley	6,962	85	99	...	...	...	...	...	31	38	...	...	...	...
Sacramento and Hope Valley, with intermediate stations	6,991	57	74	...	...	...	...	...	21	14	...	...	...	...
Placerville and Strawberry Valley	3,712	35	36	...	...	...	...	...	16	14	...	...	...	...
Strawberry Valley and Hope Valley	1,365	12	33	...	...	...	...	...	10	14	...	...	...	...
Sacramento and Placerville	1,884	40	15	...	...	...	...	...	12	9	...	...	...	...
Sacramento and Aurora	7,365	182	70	60	66	72	35	35	...	55	54	61	94	72
Sacramento and Fort Crook	3,232	179	95	41	...	...	...	...	...	39	34	...	65	60
Fort Crook and Hope Valley	3,670	185	...	...	...	...	...	...	...	57	...	...	...	...
Hope Valley and Aurora	403	63	95	...	...	...	...	...	x	...	...	...	...	...
Carson City and Hope Valley	2,335	80	...	...	...	...	...	...	...	...	...	...	...	...

From the examination of a large number of observations and computed results from daily means, I have come to the conclusion that there is no relation between the height of the barometer at the lower or upper station and the value of the differences of altitude. That is to say, the wandering from the monthly mean may be a maximum or a minimum with either a high or a low barometer.

The cause of erratic results from daily means must be attributed to the fact that the atmosphere is seldom, if ever, in a state of equilibrium, and hence the wanderings cannot be controlled by any law, and must be incident to all measurements of this kind.

OF THE VARIATIONS IN HYPSOMETRICAL RESULTS FROM MONTHLY MEANS.

When observations of the barometer and thermometer have been continued for a number of years at two stations, and the mean monthly readings are used in computing the difference of altitude between them, it has been ascertained that these computed results from observations taken in the different months differ. If we take the series of 25 years at Geneva and the Grand St. Bernard as affording us the best type series available, we find that the computed difference of altitude for the month of December and July differ by 101 feet, and that those for the different months vary by a definite law, so that when plotted they show a smooth curve. We can only ascertain what this law is by comparing results from observations taken in different latitudes, altitudes, and climates. Unfortunately, extensive series of reliable observations at high and low altitudes are seldom to be found. But I shall make use of such as I have had access to, and from which some important facts can be deduced.

Going back to the observations at Geneva and the Grand St. Bernard, the first fact of importance is, that while the

25 years' series gives a good curve, the observations taken in any one year do not, and the plotted results from monthly means of observations taken during a single year are so irregular that it would be difficult to develop from them a law in this variation. This fact shows that even with these stations, if a table of corrections were made to reduce the results taken during each of the months to the mean for the year, and if that table of corrections were applied to observations taken in any year, the correction would not with certainty be applied advantageously, though of course the chances are that they would be so applied.

It is now necessary to ascertain if this variation in hypsometrical results is peculiar to the climate of Switzerland, or whether it is applicable to other countries. The observations taken in the Sierra Nevada, though not as numerous as are desirable, at least indicate that the same general law holds good in California: but while observations in mid-winter give the least results, and those in midsummer the greatest, the range is twice as great, which may be attributed to the higher temperature of this country. The mean temperatures in the hottest and coldest months at Geneva and the Grand St. Bernard are respectively 64° and 31° at the former, and 43° and 15° at the latter. In the Sierra Nevada we have, in July, at Sacramento, $70^\circ.0$, and at the summit, about 7,000 feet high, $55^\circ.5$, and in January they are $48^\circ.9$ and $26^\circ.3$, respectively.

But unfortunately for the development of any law of practical importance that can be of use in making a table of corrections to be applied to results obtained in different months so as to reduce them to the yearly mean, the range in the variation of these results seems to depend more upon the temperatures of the stations than upon the difference of altitude between them. For example, the range between the winter and summer results, as developed from observations taken at Sacramento and Fort Churchill, is 200 feet, while the difference of altitude is about 4,200 feet, and that from observations at Sacramento and summit of the Sierra Nevada at Hope Valley, about 7,000 feet, is 118 feet. This last result, however, is from observations in the single months of July of 1860 and January of 1864.

From all that has been previously pointed out on this subject, we are certain that hypsometrical results generally give results which are considered greater in midsummer than in midwinter, but the amplitude of this variation depends so much upon the climate of the two stations that no definite rule can be given concerning it.

CONCLUDING REMARKS.

In the previous pages I have explained the method which I have recommended to be adopted in producing the best hypsometrical results, the essential points of which are to prepare the barometric observations beforehand by correcting them for the horary oscillation of the barometer, and then to use the mean daily temperature during the period in which the observations were taken, whether it be short or long.

Prof. J. D. Whitney, formerly the State geologist for the State of California, with a full knowledge of my method as explained in No. 15 of the Professional Papers of the Corps of Engineers, has adopted another method, and has explained it in a work entitled "Contributions to Barometric Hypsometry, with tables for use in California." In the first chapter the distinguished professor gives an able and learned discussion of the various forms of the barometric formula which have been used, and comes to the conclusion that no change in any of the constants is advisable. He says, in the beginning of his third chapter, that my formula, or that of Guyot (which is almost identical with it, and one or the other of which was used by him and his assistants during the geological survey of California), " is the one which leads most directly to practical results, and upon which the chief dependence is to be placed."

The third and last chapter of this work is also interesting, but the second one is the only one which explains his method of treating barometric and thermometric observations. He had obtained observations during three years at Sacramento, near the sea-level; at Colfax, on the slope of the Sierra Nevada, at an altitude of 2,414 feet; and at Summit Station, at an altitude of 6,951 feet. The altitudes were ascertained during the leveling for the railroad. The observations were taken at 7 a. m., 2 p. m., and 9 p. m. From the monthly means of the barometer and thermometer at those three hours, he ascertains how much the mean hypsometrical results at each of those hours in each month at each of the three stations differ from their altitudes as given by the level. He then forms a table of corrections to be applied to such results from observations taken in the field during those months and at those hours, and by a simple interpolation assumes corrections for the intermediate hours between 7 a. m. and 9 p. m. He uses the actual observations of the barometer and thermometer without other correction than that of reducing the barometer to 32° F. From the monthly mean pressure and temperature at each of the three hours, he deduces his table of corrections to be applied to each hour of the day and month of the year.

My method is to eliminate beforehand all causes of error, as far as possible, by first applying a correction to the barometric readings so as to get rid of the effects of the horary oscillation, and then to eliminate the effects of the horary movement by using, instead of the observed temperature, the mean daily air temperature for the period during which the barometric observations were taken.

It is evident to me that Professor Whitney's method must produce more "maximum errors" than mine; because the periods of the barometric and thermometric observations, in the field are not the same as the monthly periods of observations used by him in preparing his table of corrections. Observations in the field are usually for a short period; his table of corrections is from the monthly means. During a barometric reconnaissance, the altitudes of most of the stations on the line are approximately determined from single observations. In forming his table he has used the monthly mean of the thermometer at the three observed hours. Now it is well known that during a month the range of the thermometer during the twenty-four hours may be twice, three times, and even four times as great in one day as in another. If observations of the barometer had been taken during one day only, it might have happened that the thermometric curve (if the observations had been plotted) for that day was a maximum or a minimum, and the horary thermometric curve from the actual observations would be quite different from that obtained from the thermometric monthly mean. With regard to the barometer, its horary oscillations from day to day during a month at the same place are so nearly alike that no material error is made by adopting either the horary oscillation for the month or that for the period of observation.

Within the limits of the State of California almost every variety of climate is to be found. There is the moist and uniform climate of the coast, and the arid and tropical climate of the Mohave and Colorado deserts. There are vast plains near the sea level, and a number of mountain peaks between fourteen and fifteen thousand feet above it. When we consider the great change of temperature during twenty-four hours in a considerable portion of the State, the great variation in the range of that temperature in different days and in different localities, and the totally different character of the horary oscillations of the barometer at an upper and lower station, varying as these do with altitude, latitude, and climate, and, more than all, local peculiarities of climate, it can be easily understood that it is impossible to adopt, with good results, a table of corrections suitable to every part of such a State as California, or even to a considerable portion of it.

My method is quite as easy of application as that of Professor Whitney, except that it requires a certain amount of intelligence in the computer, who has to prepare the observations and apply the corrections before the numbers are used in computing the difference of altitude. This might be considered by some to be a disadvantage, for with his method any ignorant man, possessed of a small amount of knowledge of arithmetic, can become a barometric computer, by following certain prescribed rules. But I take it for granted that all persons engaged in barometric reconnaissances of any importance are endowed with quite enough intelligence to properly prepare barometric and thermometric observations for computation by my method.

It is of the utmost importance on a barometric reconnaissance that we should know as nearly as possible the probable maximum error, because the results are only to be fully trusted up to that limit. The method which produces the least maximum error must then be considered the best. I wish now to show, from the computed results of observations at my command, that the method adopted by Professor Whitney does actually produce more maximum error than mine.

For that purpose I have used ten days' observations at Sacramento, Placerville, Strawberry Valley, and Hope Valley, in August. Observations at those four stations are the only ones at my command which are suitable for the purpose where full hourly observations were taken. I calculated for each of the three hours during the ten days the difference of altitude by the two methods. I then made from them a table of maximum and mean errors, which is herewith submitted.

It will be seen that the amount of error in hypsometrical results is over forty per cent. more by Professor Whitney's method than by mine, and any intelligent man who has carefully studied the two methods can easily appreciate the reason.

TABLE VIII.—*Comparison of barometric results by Professor Whitney's and Colonel Williamson's methods, from observations taken at 7 a. m., 2 p. m., and 9 p. m., during ten days of August, 1860.*

SACRAMENTO AND HOPE VALLEY.

	Max. error.	Max. error.	Mean error.	Grand mean.
Whitney's method.........	+ 249. 7	— 214. 5	61. 6	6,976. 9
Williamson's method	+ 155. 7	— 144. 2	54. 5	6,961. 7
Ratio	+ 1. 60	— 1. 49	1. 13	....

SACRAMENTO AND PLACERVILLE.

Whitney's method.........	+ 37. 9	— 36. 8	13. 5	1,897. 0
Williamson's method	+ 27. 9	— 24. 6	13. 0	1,918. 4
Ratio	+ 1. 36	— 1. 50	1. 04	....

PLACERVILLE AND STRAWBERRY VALLEY.

Whitney's method	+ 194. 4	— 94. 1	42. 1	3,715. 0
Williamson's method	+ 143. 0	— 51. 5	20. 1	3,731. 6
Ratio	+ 1. 36	— 1. 66	2. 09	....

STRAWBERRY VALLEY AND HOPE VALLEY.

Whitney's method.........	+ 64. 5	— 45. 3	21. 3	1,362. 3
Williamson's method	+ 49. 2	— 41. 0	21. 0	1,368. 6
Ratio	+ 1. 31	— 1. 10	1. 01	....

4 U B